MACHINERY'S REFERENCE SERIES

EACH NUMBER IS A UNIT IN A SERIES ON ELECTRICAL AND STEAM ENGINEERING DRAWING AND MACHINE DESIGN AND SHOP PRACTISE

NUMBER 138

ELEMENTARY ALGEBRA

by

ERIK OBERG

British Library Cataloguing-in-Publication Data
A catalogue record for this book is available from
the British Library

CONTENTS

CHAPTER I

USE OF FORMULAS

In that part of mathematics known as *algebra*, letters are used to represent numbers or *quantities;* the term quantity is used to designate any number involved in a mathematical process. The use of letters or symbols, in place of the actual numbers, simplifies the solution of mathematical problems and makes the calculations easier and more rapid.

The simplest use made of letters to represent numbers is in *formulas*. A formula may be defined as a mathematical rule expressed by signs and symbols instead of in actual words. The object is to make it possible to condense into small space the information that otherwise would be imparted by long and cumbersome rules.

The symbols used in formulas, as well as in algebra in general, are mainly the letters in the alphabet, and the signs are the ordinary signs used in arithmetic, in addition to a number of signs used for special purposes in algebraic expressions only.

Letters from the Greek alphabet are frequently used to designate angles, and the Greek letter π (pi) is always used to indicate the ratio between the circumference and the diameter of a circle; π, therefore, is always, in mathematical expressions, equal to 3.1416. The Greek letters most generally used, besides π, are α (alpha), β (beta), γ (gamma), δ (delta), θ (theta), μ (mu), ϕ (phi), and ω (omega).

Knowledge of algebra is not necessary in order to make successful use of formulas of the general type, such as are found in engineering handbooks; it is only necessary to thoroughly understand the use of letters or symbols in place of numbers, and to be well versed in the methods, rules and processes of ordinary arithmetic. Knowledge of algebra becomes necessary only in cases where a general rule or formula which gives the answer to a problem directly is not available. In other words, algebra is useful in *developing* or originating a general rule or formula, but the formula can be *used* without recourse to algebraic processes. No attempt should be made to study algebra before all the methods of arithmetic are well understood.

Formulas

As mentioned, the symbols or letters used in formulas designate or "stand for" actual figures or numerical values. The figures or values for a given problem are inserted in the formula according to the requirements in each specific case. When the values are thus inserted, in place of the letters, the result or answer is obtained by ordinary arithmetical methods.

There are two reasons why a formula is preferable to a rule expressed in words. Firstly, the formula is more concise, it occupies less space, and it is possible for the eye to catch at a glance the whole meaning of the rule laid down; secondly, it is easier to remember a brief formula than a long rule, and it is, therefore, of greater value and convenience. It is not always possible to carry a handbook or reference book about, but the memory must be relied upon to store up a number of the most frequently occurring mathematical and mechanical rules.

The use of formulas can be explained most readily by actual examples. In the following, therefore, a number of simple formulas will be given, and the values will be inserted so as to show, in detail, the principles involved.

Example 1.—When the diameter of a circle is known, the circumference may be found by multiplying the diameter by 3.1416. This rule, expressed as a formula, is:

$$C = D \times 3.1416$$

in which C = circumference of circle,

D = diameter of circle.

This formula shows at a glance that the circumference of a circle is always equal to the diameter times 3.1416. The diameter may be any number of inches, feet or miles; the relation stated by the formula always exists.

Let it be required to find the circumference of a circle 22 inches in diameter. Insert 22 in the formula in place of D. Then:

$$C = 22 \times 3.1416 = 69.1152 \text{ inches.}$$

By a simple multiplication, the formula thus gives C, the circumference, equal to 69.1152 inches.

The diameter of a circle is 3.72 inches. Find the circumference.

Insert the value 3.72 in place of D in the formula. Then:

$$C = 3.72 \times 3.1416 = 11.6867 \text{ inches.}$$

Example 2.—In spur gears, the outside diameter of the gear can be found by adding 2 to the number of teeth, and dividing the sum obtained by the diametral pitch of the gear. This rule can be expressed very simply by a formula. Assume that we write D for the outside diameter of the gear, N for the number of teeth, and P for the diametral pitch. Then the formula would be:

$$D = \frac{N + 2}{P}$$

This formula reads exactly as the rule given above. It says that the outside diameter (D) of the gear equals 2 added to the number of teeth (N), and this sum divided by the pitch (P).

If the number of teeth in a gear is 16 and the diametral pitch 6, then simply put these figures in the place of N and P in the formula, and find the outside diameter as in ordinary arithmetic.

$$D = \frac{16 + 2}{6} = \frac{18}{6} = 3 \text{ inches.}$$

In another gear, the number of teeth is 96; the diametral pitch is 7. Find the outside diameter.

$$D = \frac{96 + 2}{7} = \frac{98}{7} = 14 \text{ inches.}$$

From the examples given, we may formulate the following general rule: *In formulas, each letter stands for a certain dimension or quantity; when using a formula for solving a problem, replace the letters in the formula by the figures given for a certain problem, and find the required answer as in ordinary arithmetic.*

Example 3.—The formula for the horsepower of a steam engine is as follows:

$$H = \frac{P \times L \times A \times N}{33{,}000}$$

in which H = indicated horsepower of engine;
P = mean effective pressure on piston in pounds per square inch;
L = length of piston stroke in feet;
A = area of piston in square inches;
N = number of strokes of piston per minute.

Assume that $P = 90$, $L = 2$, $A = 320$, and $N = 110$; what would be the horsepower?

If we insert the given values in the formula, we have:

$$H = \frac{90 \times 2 \times 320 \times 110}{33{,}000} = 192.$$

Omitting Multiplication Signs in Formulas

In formulas, the sign for multiplication (×) is often left out between letters the values of which are to be multiplied. Thus AB means $A \times B$, and the formula

$$\frac{P \times L \times A \times N}{33{,}000} \text{ can also be written } \frac{PLAN}{33{,}000}$$

Thus, if $A = 3$, and $B = 5$, then:

$$AB = A \times B = 3 \times 5 = 15.$$

If $A = 12$, $B = 2$, and $C = 3$, then:

$$ABC = A \times B \times C = 12 \times 2 \times 3 = 72.$$

It is only the multiplication sign (×) that can be thus left out between the symbols or letters in a formula. All other signs must be indicated the same as in arithmetic. The multiplication sign can never be left out between two figures: 35 always means thirty-five, and "three times five" must be written 3×5; but "three times A" may be written $3A$. As a general rule the figure in an expression such as "$3A$" is written first, and is known as the *coefficient* of A. If the letter is written first, the multiplication sign is not left out, but the expression is written "$A \times 3$."

Use of Parentheses

Parentheses () or brackets [] in a formula, or in an algebraic expression in general, indicate that the expression inside the parentheses or brackets should be considered as one single symbol, or in other words, that the calculation inside the parentheses or brackets should be carried out by itself, before other calcnlations are carried out.

Examples:

$6 \times (8 + 3) = 6 \times 11 = 66.$

$5 \times (16 - 14) + 3\ (2.25 - 1.75) = 5 \times 2 + 3 \times 0.5 = 10 + 1.5 = 11.5.$

In the last example it will be seen that 5 is multiplied by 2 and 3 by 0.5, and then the products of these two multiplications are added. From the order of the numbers $5 \times 2 + 3 \times 0.5$, one might have assumed that the calculation should have been carried out as follows: 5 times $2 = 10$, plus $3 = 13$, times $0.5 = 6.5$. This latter procedure, however, is not correct, as explained in the following paragraph.

Order of Operations

When several numbers or expressions are connected by the signs $+$, $-$, $\times$ and $\div$, the operations are carried out in the order written, except that *all multiplications should be carried out before the other operations.* The reason for this is that numbers connected by a multiplication sign are only factors of the product thus indicated, which product should be considered by itself as one number. Divisions should be carried out before additions and subtractions, if the division is indicated in the same line with these other processes and if there are no parentheses indicating that other operations must first be worked out.

Examples:

$5 \times 6 + 4 - 6 \times 4 = 30 + 4 - 24 = 34 - 24 = 10.$

$5 + 3 \times 2 = 5 + 6 = 11.$

$100 \div 2 \times 5 = 100 \div 10 = 10.$

$3.5 + 16.5 \div 3 - 1.75 = 3.5 + 5.5 - 1.75 = 7.25.$

But $5 \times (6 + 4) - 6 \times 4 = 5 \times 10 - 24 = 50 - 24 = 26.$

$(5 + 3) \times 2 = 8 \times 2 = 16.$

$(100 \div 2) \times 5 = 50 \times 5 = 250.$

$(3.5 + 16.5) \div (3 - 1.75) = 20 \div 1.25 = 16.$

Examples for Practice

(1). Find the value of P in the formula:

$$P = \frac{R + N}{B}$$

if $R = 22$: $N = 2$; and $B = 16$. Answer: $1\frac{1}{2}$.

(2). Find the value of S in the formula:

$$S = \frac{AB + C}{CD}$$

if $A = 124$; $B = 0.25$; $C = 2$; and $D = 66$. Answer: 0.25.

(3). Find the value of N in the formula:

$$N = \frac{2R + 3A}{A + B}$$

if $A = 27$; $B = 6$; $R = 9$. Answer: 3.

(4). Find the value of F in the formula:

$$F = \frac{(A + B) \times (B + C) \times (C + D)}{A\,B\,C\,D}$$

if $A = 1/2$; $B = 1/3$; $C = 1/4$; $D = 1/5$. Answer: 26¼.

Exponents

The square of a number is the product of that number multiplied by itself. The square of 2 is $2 \times 2 = 4$, and the square of 10 is $10 \times 10 = 100$; similarly the square of 177 is $177 \times 177 = 31{,}329$. Instead of writing 4×4 for the square of 4, it is often written 4^2 which is read *four square*, and means that 4 is multiplied by 4. In the same way 128^2 means 128×128. The small figure (^{2}) in these expressions is called *exponent*.

The cube of a number is the product obtained if the number itself is repeated as a factor three times. The cnbe of 2 is $2 \times 2 \times 2 = 8$, and the cnbe of 12 is $12 \times 12 \times 12 = 1728$. Instead of writing $2 \times 2 \times 2$ for the cube of 2, it is often written 2^3, which is read *two cube*. In the same way 128^3 means $128 \times 128 \times 128$. The small figure (^{3}) in these expressions is the *exponent*, the same as in the case of the figure (^{2}) indicating the square of a number. An expression of the form 18^3 may also be read the "third power of 18."

In the same way as we write $2 \times 2 = 2^2$, and $2 \times 2 \times 2 = 2^3$, we can write $2^4 = 2 \times 2 \times 2 \times 2$. Similarly, the expression 2^5 means that 2 is repeated as a factor five times, or:

$$2^5 = 2 \times 2 \times 2 \times 2 \times 2 = 32.$$

The expression 2^4 is read "the fourth power of 2," and the expression 6^5, "the fifth power of 6." Thus, the expression "seventh power" would mean that the exponent equals 7.

From these examples it is evident that the object of the exponent is to indicate how many times the number to which the exponent is affixed is to be taken as a factor.

Exponents may be affixed to letters or symbols as well as to figures. For example, a^5 indicates that a is to be taken as a factor five times, or:

$$a^5 = a\,a\,a\,a\,a = a \times a \times a \times a \times a.$$

In an expression of the form $7a^3$, the expouent applies only to the symbol a to which it is affixed, and not to the coefficient "7." If $a = 3$, then:

$$7a^3 = 7 \times 3^3 = 7 \times 3 \times 3 \times 3 = 189.$$

The exponent can, however, be made to apply to the coefficient also, by the use of parentheses enclosing the expression to which the exponent applies:

$$(7a)^3 = (7 \times 3)^3 = (21)^3 = 21 \times 21 \times 21 = 9261.$$

If the exponent applies only to the coefficient, the expression would be written "7^3a." The value, if $a = 3$, would be:

$$7^3a = 7 \times 7 \times 7 \times 3 = 1029.$$

The meaning of exponents will be made still clearer by the following examples:

$$a^2b^3 = a \times a \times b \times b \times b.$$
$$(ab)^2 = a^2b^2 = a \times a \times b \times b.$$
$$3(ab)^3 = 3a^3b^3 = 3 \times a \times a \times a \times b \times b \times b.$$

Roots

The square root of a number is that number which, when multiplied by itself, will give a product equal to the given number. Thus, the square root of 4 is 2, because 2 multiplied by itself gives 4. The square root of 25 is 5; of 36, 6, etc. We may say that the square root is the reverse of the square, so that if the square of 24 is 576, then the square root of 576 is 24. The mathematical sign for the square root is $\sqrt[2]{\ }$, but the *index figure* (^{2}) is generally left out, making the square-root sign simply $\sqrt{\ }$, thus:

$\sqrt{4} = 2$ (the square root of four equals two),

$\sqrt{100} = 10$ (the square root of one hundred equals ten).

The operation of finding the square root of a given number is called *extracting* the square root. Squares and square roots as well as cubes and cube roots of all numbers up to 1000 (sometimes up to 1600 and in MACHINERY'S HANDBOOK up to 2000) are generally given in all standard handbooks.

In the same way as square root means the reverse of square, so cube root means the reverse of cube; that is, the cube root of a given number is the number which, if repeated as a factor three times, would give the number given. Thus the cube root of 27 is 3, because $3 \times 3 \times 3 = 27$. If the cube of 15 is 3375, then the cube root of 3375 is, of course, 15. The mathematical sign for the cube root is $\sqrt[3]{\ }$, thus:

$\sqrt[3]{64} = 4$ (the cube root of sixty-four equals four),

$\sqrt[3]{4096} = 16$ (the cube root of four thousand ninety-six equals sixteen).

In the same way as we may say that the square root means the reverse of square, and the cube root the reverse of cube, so we may say that the fourth root is the reverse of the fourth power; that is, if we require a number, which when repeated as a factor four times will give as a product a given number, we must obtain the fourth root, or $\sqrt[4]{\ }$

Thus $\sqrt[4]{81} = 3$, because $3 \times 3 \times 3 \times 3 = 81$.

The fifth root is written $\sqrt[5]{\ }$; and, as an example:

$$\sqrt[5]{32} = 2, \text{ because } 2 \times 2 \times 2 \times 2 \times 2 = 32.$$

Symbols may, of course, be used as well as figures. For example:

$$\sqrt[3]{ab}; \quad \sqrt[5]{n}; \quad \sqrt{a+b}; \quad \sqrt[n]{a}; \quad \sqrt[n]{17}.$$

In the same way as $3^2 = 3 \times 3$, so $a^2 = a \times a$; and as $\sqrt{9} = 3$, so also $\sqrt{a^2} = a$, because $a \times a = a^2$.

The principles of roots applied to algebraic expressions are shown by the examples below:

$$\sqrt{a^2} = a; \quad \sqrt[3]{a^3} = a; \quad \sqrt{a^2b^2} = ab.$$
$$\sqrt{4a^2} = 2a; \quad \sqrt[3]{27a^3} = 3a; \quad \sqrt{16a^2b^2} = 4ab.$$

Expressions, such as $\sqrt{a}$, cannot be further simplified, because the root cannot be extracted from a quantity consisting of a letter without an exponent, except by introducing *fractional exponents*, which belongs to a more advanced stage of algebraic study.

The square and cube roots of numbers may be extracted by methods known from arithmetic. (See MACHINERY'S Reference Book No. 52, "Advanced Shop Arithmetic for the Machinist.") In practice, however, the tables of squares and cubes, and square roots and cube roots, given in standard handbooks, are used to avoid the time-consuming and cumbersome methods otherwise necessary to employ.

Examples Involving Roots and Exponents

Example 1.—Find the value of A in the formula:

$$A = \frac{\sqrt{B} \times C}{D}$$

Assume $B = 36$; $C = 3.5$; $D = 10.5$.

By inserting these values in the formula, we have:

$$A = \frac{\sqrt{36} \times 3.5}{10.5} = \frac{6 \times 3.5}{10.5} = \frac{21}{10.5} = 2.$$

Example 2.—Find the value of A in the formula:

$$A = \frac{B^2 + C^2}{D^2}$$

if $B = 10$; $C = 14$; and $D = 4$.

$$A = \frac{10^2 + 14^2}{4^2} = \frac{10 \times 10 + 14 \times 14}{4 \times 4} = \frac{100 + 196}{16} = 18.5.$$

Example 3.—Find the value of A in the formula:

$$A = \sqrt{B^2 + C^2}, \text{ if } B = 8 \text{ and } C = 6.$$

If we insert the given values in the formula, we have:

$$A = \sqrt{8^2 + 6^2} = \sqrt{8 \times 8 + 6 \times 6} = \sqrt{64 + 36} = \sqrt{100} = 10.$$

The examples given indicate the principles involved in the use of formulas, and show also how easily formulas may be employed by anyone who has a general understanding of arithmetic.

CHAPTER II

USE OF EQUATIONS IN SOLVING PROBLEMS

An *equation* is a statement of equality between two expressions. Thus, $5x = 105$, is an equation. Equations are used for the solution of mathematical problems. When a problem is presented, there is always one or more quantities to be found as an answer to the problem. These quantities are called *unknown* quantities. If there is only one unknown quantity in a problem, it is generally designated by the letter x in the equation used as an aid in the solution of the problem. If there is more than one unknown quantity, the others are designated by letters also selected at the end of the alphabet, as y, z, u, t, etc.

Different Types of Equations

An equation is said to be of the *first degree*, if it contains the unknown in the first power only; that is, if the unknown in the equation has no exponent. For example, $3x = 9$, is an equation of the first degree, because x has no exponent. Strictly speaking, the exponent is (^{1}), but (^{1}) is never written out as an exponent. It is well to always bear in mind, however, that x really means x^1. In the same way, note that a letter without a numerical coefficient, strictly speaking, has a coefficient equal to 1; so that, for example, x actually equals $1x$, or even $1x^1$, although *exponents and coefficients equal to 1 are never written out.*

An equation which contains the unknown in the second, or first and second, but no higher, power, is called a *quadratic* equation. Example: $x^2 + 3x = 18$.

An equation which contains the unknown in the third power is called a *cubic* equation. Example: $x^3 + 3x^2 + x = 22$.

For the present, we shall confine our attention to equations of the first degree, or *simple* equations, and to the problems that can be solved by means of them. The use of equations involves the use of letters or symbols; and calculations in which symbols are used to represent numbers or quantities require a knowledge of algebra. In the following the use of equations and the elements of algebra will be briefly treated simultaneously, by examples showing the application to practical problems. Many of these problems are so simple that they can be solved readily by arithmetic without recourse to equations; but these simple examples are here used to show in a clear manner the principles involved in the use of equations. By this step by step method, a working knowledge and a conception of the importance of the subject can most easily be acquired. Later, it will be necessary to study the laws and methods of algebra in greater detail, in order to fully grasp the subject.

Use of Simple Equations

Problem 1.—The cost of 9 pounds of tin is $2.61. What is the price of tin per pound?

In solving any problem by means of an equation, first determine what is the unknown quantity, and call this quantity x. In the problem given, the price per pound is x. Then insert this x into an equation, making use of the information given in the problem. In this case, we have:

$$9x = 261.$$

It is evident that if the quantity to the left of the equal sign ($9x$) equals the quantity to the right of the equal sign (261), then these quantities will also be equal if both are divided by the same nnmber. Hence,

$$\frac{9x}{9} = \frac{261}{9}$$

By cancellation, and division, this becomes:

$$x = 29.$$

This is the answer to our problem. The price of one pound of tin is 29 cents.

Problem 2.—It is known that 3/7 of the total capacity of a water tank is 39 gallons. Find the total capacity.

Assume that the total capacity is x. Then:

$$3/7x = 39.$$

In the same way as we can divide both sides of an equation by the same number, without disturbing the condition of equality, so we can also multiply both sides with the same number, and still have the two sides or members equal. Hence,

$$7/3 \times 3/7x = 7/3 \times 39.$$

By carrying ont the arithmetical work:

$$x = 91.$$

The capacity of the water tank is 91 gallons.

From the two examples given it will be seen that the object of the division in Problem (1) and of the multiplication in Problem (2) was to obtain the unknown quantity x in a form where it would have a coefficient equal to 1. When the unknown x, with a coefficient equal to 1, is on one side of the equal sign, and only known numbers or quantities on the other, the problem is solved. The only difficulty met with in equations is to simplify the equation to this form.

Problem 3.—A man working 54 hours a week pays, at the end of the week, $7.30 out of his week's pay; he has $8.90 left. How much does he earn per hour.

Let x cents be the earnings per hour; then $54x$ are the total earnings per week, in cents. Out of this 730 cents are paid out, and the remainder equals 890 cents. Hence,

$$54x - 730 = 890.$$

To solve this equation, all the known quantities must be transposed to the right-hand side. To do this, add 730 to both members of the equation; this will not change the condition of equality.

$$54x - 730 + 730 = 890 + 730.$$

By simplifying on each side of the equal sign:

$$54x = 1620.$$

Now divide both sides by 54, the coefficient of x:

$$\frac{54x}{54} = \frac{1620}{54}.$$

$$x = 30 \text{ cents per hour.}$$

Transposition

It will be seen that in the equation in Problem (3), the effect of adding 730 to both sides of the equation was to change its form from

$$54x - 730 = 890$$

to the form

$$54x = 890 + 730.$$

This involves an important rule for the solution of equations: *Any independent term may be transposed from one side of the equal sign to the other by simply changing its sign;* that is, + on one side of the equal sign becomes — on the other side, and *vice versa*. By *independent term* is meant one not tied to the other terms by signs or arrangements implying multiplication or division.

The following examples will show the application of the rule of transposition to equations.

Example 1.

$$22x - 11 = 15x + 10;$$
$$22x - 15x = 10 + 11;$$
$$7x = 21;$$
$$x = 3.$$

A term, as $15x$, to the right in the first line, not preceded by any sign, is always assumed to be preceded by a + sign, or to be *positive*. Hence, when this term was transposed to the left-hand side, it became a *negative* term, or preceded by a — sign.

Example 2.

$$12x - 93 - (3x + 1) = 12.$$

When a — sign precedes a parenthesis, the parenthesis may be removed, if the signs of all the terms within the parenthesis are changed. Hence,

$$12x - 93 - 3x - 1 = 12,$$

and by transposing all known terms to the right side of the equation:

$$12x - 3x = 12 + 93 + 1$$
$$9x = 106$$
$$x = \frac{106}{9} = 11\ 7/9$$

General Rule for Solving Equations of the First Degree

We may now formulate the following rule for the solving of equations of the simple form so far considered:

Transpose all the terms containing the unknown quantity x to one side of the equal sign, and all the other terms to the other side. Combine and simplify the expressions on either side as far as possible, and then divide both sides by the coefficient of the unknown x.

Examples for Practice

The examples in the following are worked out in order to clearly show the student the method of procedure.

1. If it takes 18 days to assemble 4 machines, how many days would be required to assemble 14 machines?

$x =$ the number of days required.

The problem is one of proportion, and

$$x : 14 = 18 : 4, \text{ or}$$

$$\frac{x}{14} = \frac{18}{4}$$

Multiply both sides by 14 and cancel; this will give the equation a form in which x has a coefficient equal to 1:

$$\frac{x}{14} \times 14 = \frac{18}{4} \times 14$$

$$x = 63 \text{ days.}$$

2. Two trains start simultaneously from two terminals 360 miles apart, traveling towards each other. One train averages 50 miles an honr, the other 30 miles an hour. How soon do they meet?

The trains will meet after x honrs. The faster train will then have traveled $50x$ miles, the slower, $30x$ miles. The total distance traveled by the two trains is equal to the distance between terminals. Hence,

$$50x + 30x = 360$$

$$80x = 360$$

$$x = \frac{360}{80} = 4\frac{1}{2} \text{ hours.}$$

3. One thousand dollars are to be divided between fonr persons, A, B, C and D, so that A gets $50 more than B, and C gets $150 more than D. The shares of B and D are to be equal. Find the share of each.

B and D will have x dollars each.

A will receive $x + 50$ dollars.

C will receive $x + 150$ dollars.

The sum of the four shares is 1000 dollars. Hence,

$$x + x + x + 50 + x + 150 = 1000.$$

$$4x + 200 = 1000.$$

$$4x = 1000 - 200.$$

$$4x = 800.$$

$$x = 200.$$

A's share is $200 + 50 = 250$; B's, 200; C's, $200 + 150 = 350$; and D's 200 dollars.

CHAPTER III

PRINCIPLES OF ALGEBRA

Positive and Negative Quantities

On the thermometer scale, as is well known, the graduations extend upward from zero, the degrees being numbered 1, 2, 3, etc. Graduations also extend downward and are numbered in the same way: 1, 2, 3, etc. The degrees on the scale extending upward from the zero point may be called *positive* and preceded by a plus sign, so that, for instance, + 5 degrees means 5 degrees above zero. The degrees below zero may be called *negative* and may be preceded by a minus sign, so that — 5 degrees means 5 degrees below zero.

The ordinary numbers may also be considered positive and negative in the same way as the graduations on a thermometer scale. When we count 1, 2, 3, etc., we refer to the numbers that are larger than 0 (corresponding to the degrees *above* the zero point), and these numbers are called positive numbers. We can conceive, however, of numbers extending in the other direction of 0; numbers that are, in fact, less than 0 (corresponding to the degrees below the zero point on the thermometer scale). As these numbers must be expressed by the same figures as the positive numbers, they are designated by a minus sign placed before them. For example, — 3 means a number that is as much less than, or beyond 0 in the negative direction as 3 (or, as it might be written, + 3) is larger than 0 in the positive direction.

A negative value should always be enclosed within a parenthesis whenever it is written in line with other numbers; for example:

$$17 + (-13) - 3 \times (-0.76).$$

In this example — 13 and — 0.76 are negative numbers, and by enclosing the whole number, minus sign and all, in a parenthesis, it is shown that the minus sign is part of the number itself, indicating its negative value.

It must be understood that when we say 7 — 4, then 4 is not a negative number, although it is preceded by a minus sign. In this case the minus sign is simply the sign of subtraction, indicating that 4 is to be subtracted from 7. But 4 is still a positive number or a number that is larger than 0.

It now being clearly understood that positive numbers are all ordinary numbers greater than 0, while negative numbers are conceived of as less than 0, and preceded by a minus sign which is a part of the number itself, we can give the following rules for calculations with negative numbers. The application of the expressions positive and negative to algebraic terms involving symbols instead of figures follows the same rules, as the examples indicate.

A negative number can be added to a positive number by subtracting its numerical value from the positive number.

Examples:

$4 + (-3) = 4 - 3 = 1.$

$16 + (-7) + (-6) = 16 - 7 - 6 = 3.$

$327 + (-0.5) - 212 = 327 - 0.5 - 212 = 114.5.$

In the last example 212 is not a negative number, because there is no parenthesis indicating that the minus sign is a part of the number itself. The minus sign, then, indicates only that 212 is to be subtracted in the ordinary manner.

$$A + (-B) = A - B.$$
$$A + (-B - C) = A - B - C.$$

The last example shows the application of an important rule relating to the use of parenthesis in algebra. *If a parenthesis is preceded by a + sign, it may be removed, if the terms inside the parenthesis retain their signs.*

A negative number can be subtracted from a positive number by adding its numerical value to the positive number.

Examples:

$4 - (-3) = 4 + 3 = 7.$

$16 - (-7) = 16 + 7 = 23.$

$327 - (-0.5) - 212 = 327 + 0.5 - 212 = 115.5.$

In the last example, note that 212 is subtracted, because the minus sign in front of it does not indicate that 212 is a negative number.

$$A - (-B) = A + B.$$
$$A - (-B - C) = A + B + C.$$

The last two examples are an application of the rule that a *parenthesis preceded by a — sign may be removed if the signs preceding each of the terms inside the parenthesis have their signs changed* (+ changed to —, and — to +). Multiplication and division signs are not affected.

Examples:

$A - (-B + C + D) = A + B - C + D.$

$A - (B - C + D \times E) = A - B + C - D \times E.$

Remember that B in the last example, with no sign in front of it inside of the parenthesis, is considered as preceded by a + sign. Hence this + sign is changed to — when removing the parenthesis.

As an illustration of the method used when subtracting a negative number from a positive one, assume that we are required to find how many degrees difference there is between 37 degrees above zero and 24 degrees below; this latter may be written (—24). The difference between the two numbers of degrees mentioned is then:

$$37 - (-24) = 37 + 24 = 61.$$

A little thonght makes it obvious that this result is right, and the example shows that the rule given is based on correct reasoning.

When a positive number is multiplied or divided by a negative number, multiply or divide the numerical values as usual; but the product or quotient, respectively, becomes negative. The same rule holds true if a negative number is multiplied or divided by a positive number.

Examples:

$4 \times (-3) = -12.$ $(-3) \times 4 = -12.$

$\frac{15}{-3} = -5.$ $\frac{-15}{3} = -5.$

$A \times (-B) = -AB.$

$\frac{A}{-B} = -\frac{A}{B}.$ $\frac{-A}{B} = -\frac{A}{B}.$

When two negative numbers are multiplied by each other, the product is positive. When a negative number is divided by another negative number the quotient is positive.

Examples:

$(-4) \times (-3) = 12.$ $\frac{-4}{-3} = 1.333.$

$(-A) \times (-B) = AB.$ $\frac{-A}{-B} = \frac{A}{B}.$

If, in a subtraction, the number to be subtracted is larger than the number from which it is to be subtracted, the calculation can be carried out by subtracting the smaller number from the larger, and indicating that the remainder is negative.

Examples:

$3 - 5 = -(5 - 3) = -2.$

In this example 5 cannot, of course, be subtracted from 3, but the numbers are reversed, 3 being subtracted from 5, and the remainder indicated as being negative by placing a minus sign before it.

$227 - 375 = -(375 - 227) = -148.$

$7a - 9a = -2a.$

The examples given, if carefully studied, will enable the student to carry out calculations with negative numbers and quantities when required. The rules given are highly important in all algebraic calculations, and must be committed to memory.

Addition and Subtraction in Algebra

Addition in algebra is defined as the process of finding the sum which equals the combined value of the quantities added.

Subtraction in algebra is the process of finding the difference between two quantities. As a rule, addition and subtraction in algebra are most conveniently treated together under one head.

Only *like terms* can be added or subtracted. By like terms are meant those which differ only in their numerical coefficients. For example, $9xy$ and $\frac{1}{2}xy$ are like terms; but $3x$ and $3x^3$ are unlike terms, because here the terms differ with respect to exponents.

$7a^2$, $6a^2$, $\frac{3}{4}a^2$, and a^2 are all like terms. $5a^2$, $5a$, $5ab$, $5b$, and b^2 are all unlike terms.

Unlike terms cannot be added or subtracted, but the addition or subtraction may be *indicated* by placing + or — signs, respectively, between the terms.

Examples of adding and subtracting like terms are given below:

(1). $a + a = 2a$.
(2). $x - x = 0$.
(3). $2a + a = 3a$.
(4). $2x - x = x$.
(5). $3a + 2a = 5a$.
(6). $5x - 2x = 3x$.
(7). $3a + 6a + 2a = 11a$.
(8). $5x + 9x + x = 15x$.
(9). $3xy + 5xy = 8xy$.
(10). $ax + 5ax = 6ax$.
(11). $19abc - abc = 18abc$.
(12). $2\ xyz + xyz = 3xyz$.

Examples of addition or subtraction of unlike terms, which cannot be added directly, are given below. Add all like terms where possible.

(1). $2xy + xy + a = 3xy + a$.
(2). $3a + b + 2a = 5a + b$.
(3). $2a + a - b = 3a - b$.
(4). $5x - y + 2x = 7x - y$.
(5). $3xy + 3x + 3y + 2xy + 2x = 5xy + 5x + 3y$.

If any terms are enclosed in parentheses, these are removed, the signs of the quantities within the parentheses being changed if the parentheses is preceded by a — sign, according to the rules given under the head "Positive and Negative Quantities."

Examples:

(1). $a + (b - c) = a + b - c$.
(2). $a - (b - c) = a - b + c$.
(3). $2a - (a + b) = 2a - a - b = a - b$.
(4). $5a + (3a + b) = 5a + 3a + b = 8a + b$.
(5). $3xy - (5xy - 2y) = 3xy - 5xy + 2y = -2xy + 2y$.

The general rules for addition and subtraction of algebraic quantities may be stated as follows:

First remove all parentheses, changing the signs if required by the rules previously given.

Add the coefficients of all positive like terms.

Next, add the coefficients of all negative like terms.

Subtract the less of these two sums from the greater, and prefix the sign of the greater sum to the result; then annex the like symbols to this coefficient.

Examples:

(1). $3xy - 5xy + 3xy + 2xy - 2xy = 8xy - 7xy = xy$.
(2). $2ab - (5ab - 6ab + ab) = 2ab - 5ab + 6ab - ab = 8ab - 6ab$
$= 2ab$.
(3). $-2ax + 5ax - 3ax + 1 = 5ax - 5ax + 1 = 1$.
(4). $5a + 6b + 2a - 3b - 2c = 7a + 3b - 2c$.

(5). $6x - (5y - x) + 2y - (x + 5y) =$
$6x - 5y + x + 2y - x - 5y = 6x - 8y.$

(6). $6x - \{y - [7x - 4] + (x - y)\} =$
$6x - y + [7x - 4] - (x - y) =$
$6x - y + 7x - 4 - x + y = 12x - 4.$

(7). $-(2xy - 2x - 2y) = -2xy + 2x + 2y$, or as it is generally written, $2x + 2y - 2xy$.

(8). $-[-x - (-y + x)] = +x + (-y + x) = x - y + x$
$= 2x - y.$

Multiplication

The first rule for multiplication in algebra is as follows: *To find the product of two or more quantities, multiply together their coefficients, and prefix this product to the quantities expressed by symbols.* Remember that expressions such as $5ab$ imply the multiplication $5 \times a \times b$.

Examples:

(1). $5x \times 3y = 5 \times x \times 3 \times y = 15 \times x \times y = 15xy.$
(2). $3a \times 4b \times 5c = 3 \times 4 \times 5 \times a \times b \times c = 60abc.$
(3). $6ab \times 2c = 12abc.$
(4). $xy \times 3z = 3xyz.$

The second rule for multiplication in algebra is: *The sum of the exponents of the letters in the factors to be multiplied, equals the exponent of the same letters in the product.* Remember that a has, in fact, the exponent (^{1}), although we never write it a^1, but simply drop the exponent when it is (^{1}).

Examples:

(1). $a^2 \times a^3 = (a \times a) \times (a \times a \times a) = a^5.$
(2). $a \times a^2 = a^3.$
(3). $a^2b \times a = a^2ab = a^3b.$
(4). $a^2b \times ab^2 = a^{2+1} b^{1+2} = a^3b^3.$
(5). $a \times a^2b \times ab^2 \times b^3 = a^4b^6.$
(6). $6ab^2 \times 3a^2b^2 = 18a^3b^4.$

The last example illustrates the simple manner in which we may express the two rules given: *The coefficients are multiplied, and the exponents are added together.*

Referring to the rules given for positive and negative numbers, we may formulate the third rule for multiplication as follows: *Two positive (+) factors give a positive (+) product; two negative (—) factors give a positive (+) product; but one positive (+) and one negative (—) factor give a negative (—) product.* This rule is often stated more briefly as follows: *Like signs produce plus, and unlike signs produce minus in the product.*

Examples:

(1). $2xyz \times 4x^2 = 8x^3yz.$
(2). $-3ab \times -6ab = +18a^2b^2.$
(3). $3ab \times -6ab = -18a^2b^2.$
(4). $-3ab \times 6ab = -18a^2b^2.$

When three or more quantities that are not all positive are to be multiplied, the safest method is to multiply two' factors at a time, until all have been multiplied. In this way errors in the sign of the product are avoided.

Examples:

(1). $2ab \times -3b \times -4a = -6ab^2 \times -4a = +24a^2b^2$.

(2). $-a \times -b \times -c \times -d = ab \times -c \times -d = -abc \times -d = +abcd$.

(3). $16a^2b^2 \times -16a^2b^2 \times -a = -256a^4b^4 \times -a = 256a^5b^4$.

Expressions within parentheses may be considered as single letters, especially if they are higher than the first power.

(4). $(a-b)^2 \times (a-b) = (a-b)^3$.

(5). $(a+b) \times -(a+b)^2 = -(a+b)^3$.

If an expression consisting of several terms is enclosed in parentheses, and this expression is to be multiplied by another term, the multiplication sign is often omitted, the same as between individual letters. For example, $(x - y + z)\,a = (x - y + z) \times a$.

When one factor consists of several terms enclosed in parentheses, and the other factor is a single term, *multiply each of the terms in the parentheses by the single term.* Remember that in this case also like signs produce plus and unlike signs minus.

Examples:

(1). $(x + y - t)\,a = ax + ay - at$.

(2). $b\,(ab - ab^2 + a^2) = ab^2 - ab^3 + a^2b$.

(3). $-ab\,(a^2 - 2ab + b^2) = -a^3b + 2a^2b^2 - ab^3$.

(4). $4x\,(x - 3y) = 4x^2 - 12xy$.

(5). $3ab\,(2a - 3ab - 21c) = 6a^2b - 9a^2b^2 - 63abc$.

(6). $-\frac{1}{2}x\,(xy - x^2 + 3y^2 - 6xy^2) = -\frac{1}{2}x^2y + \frac{1}{2}x^3 - 1\frac{1}{2}xy^2 + 3x^2y^2$.

If both factors consist of a number of terms, *multiply each of the terms of one factor by each term of the other factor.* Simplify the expression thus obtained by adding the partial products algebraically, if possible.

Examples:

(1). $(a + b)(a + b) = a^2 + ab + ab + b^2 = a^2 + 2ab + b^2$.

(2). $(a + b)(a - b) = a^2 + ab - ab - b^2 = a^2 - b^2$.

(3). $(a - b)(a - b) = a^2 - ab - ab + b^2 = a^2 - 2ab + b^2$.

(4). $(3x - 2y)(5x + 4y) = 15x^2 - 10xy + 12xy - 8y^2 = 15x^2 + 2xy - 8y^2$.

(5). $(2a - 3b + 4c)(a + b + c) = 2a^2 - 3ab + 4ac + 2ab - 3b^2 + 4bc + 2ac - 3bc + 4c^2 = 2a^2 - 3b^2 + 4c^2 - ab + 6ac + bc$.

(6). $(2a - 2b)^2 = (2a - 2b)(2a - 2b) = 4a^2 - 4ab - 4ab + 4b^2 = 4a^2 - 8ab + 4b^2$.

Division

The following are the fundamental rules for division in algebra:

1. *The coefficient of the quotient equals the coefficient of the dividend divided by the coefficient of the divisor.*

2. *The exponent of a letter in the quotient equals the exponent of the same letter in the dividend, minus the exponent of the letter in the divisor.*

3. *Like signs in dividend and divisor produce a positive (+) quotient. Unlike signs produce a negative (—) quotient.*

Examples:

(1). $6ab \div 3 = 2ab$.

(2). $6a^2b^2 \div 3ab = 2a^{2-1}\,b^{2-1} = 2ab$.

(3). $15a^5 \div 5a^2 = 3a^3$.

(4). $-a^4 \div a^2 = -a^2$.

(5). $a^4 \div -a = -a^3$.

(6). $-a^5 \div -a^4 = +a^{5-4} = a$.

(7). $a^3 \div a^3 = a^{3-3} = a^0 = 1$.

We know from the rnles of arithmetic that when dividend and divisor are alike, the quotient equals 1; this last example, therefore, indicates that any quantity the exponent of which is (^{0}) is equal to 1.

(8). $\dfrac{12x^5y^4t^3}{-6x^2yt^3} = -2x^3y^3$.

It will be remembered from the rules for arithmetic that a fraction line is equivalent to a division sign.

When the dividend consists of more than one term, bnt the divisor of one term only, divide each of the terms of the dividend by the divisor.

Examples:

(1). $(6x^2y^2 - 4x^2y + 2xy^2) \div 2xy = 3xy - 2x + y$.

(2). $(-39a^3b^2 + 18a^2b^2 - 27ab^2) \div -3ab^2 = 13a^2 - 6a + 9$.

(3). $\dfrac{m^4n^3 - m^3n^2 + 4m^2n^2}{4mn^2} = \frac{1}{4}m^3n - \frac{1}{4}m^2 + m$.

When both the dividend and divisor consist of several terms, the division can be worked out only if the divisor is a factor of the dividend, which in practical problems is not often the case. The method followed, while of little practical value, is indicated below. The rules of procedure are:

1. Arrange dividend and divisor according to the power of some one letter.

2. Divide the first term of the dividend by the first term of the divisor. This gives the first term of the quotient.

3. Mnltiply all the terms of the divisor by the first term of the quotient, just found; subtract this product from the dividend.

4. The remainder is regarded as a new dividend. Its first term is divided by the first term of the divisor, to obtain the second term in the quotient.

5. Multiply all the terms of the divisor by the second term of the quotient; snbtract this product from the first remainder.

6. Continue this procedure until the remainder becomes 0. If no such remainder is obtained, but instead a remainder is found the first term of which cannot be divided by the first term of the divisor, then the division cannot be carried out, because in that case the divisor is not a factor of the dividend.

In this calculation, arrange the terms as below:

$$\text{Dividend} \,\big|\, \frac{\text{divisor}}{\text{quotient}}$$

As an example divide:

$$(x^2 - 72 + x) \div (9 + x)$$

Arrange according to the power of x, and write out as shown above:

$$\begin{array}{r|l} x^2 + x - 72 & x + 9 \\ \text{Subtract } x^2 + 9x & \overline{x - 8} \\ \hline \text{First remaiuder } -8x - 72 & \\ \text{Subtract} -8x - 72 & \\ \hline 0 \quad 0 & \end{array}$$

Factoring

Factoring, in algebra, is a most important operation, because a great number of the calculations with letters are carried ont merely by factoring and cancellation. The rules for cancellation are identical with those in arithmetic, like factors being cancelled. For factoring in algebra, however, a number of different rules must be applied.

1. Any simple algebraic quantity of more than one factor can always be resolved into its factors. For example $9a^2x^2 = 9 \times ax \times ax$, or $9 \times a^2 \times x^2$, or $9 \times a \times a \times x \times x$.

2. When an expression consists of several terms, all of which have a common factor, the expression may be resolved into two factors by dividing each term by the common factor. It is evident that the factoring can be proved by multiplying together the factors, which will then give the given quantity as a product.

Examples:

(1). Factor $27x^2y - 18xy^2 + 12x^2y^2$.

The common factor is $3xy$. Dividing all the terms by this we have as the factors:

$$3xy\ (9x - 6y + 4xy).$$

(2). $16m^2n^2 - 12mn = 4mn\ (4mn - 3)$.

(3). $2abc - 4ab - 6bc = 2b\,(ac - 2a - 3c)$.

(4). $3a^3 + 3a^2 + 3a = 3a(a^2 + a + 1)$.

3. There are a number of algebraic expressions the factors of which should be committed to memory, becanse of the frequency with which they occur in calculations. The most common of these are:

(1). $a^2 + 2ab + b^2 = (a + b)(a + b)$.

(2). $a^2 - 2ab + b^2 = (a - b)(a - b)$.

(3). $a^2 - b^2 \qquad = (a + b)(a - b)$.

Note that in Examples (1) and (2) the first and last terms are the *squares* of terms a and b and the second or middle term equals *twice the product* of terms a and b. The sign of the middle term depends on the sign between a and b in the factors. In Example (3), where the factors are $(a + b)$ and $(a - b)$, there is no middle term, and b^2 is negative. The following expressions can be resolved into factors in a similar manner:

$$4a^2 + 8ab + 4b^2 = (2a + 2b)(2a + 2b).$$
$$4a^2 - 8ab + 4b^2 = (2a - 2b)(2a - 2b).$$
$$4a^2 - 4b^2 = (2a + 2b)(2a - 2b).$$

and still further:

$$36m^2 + 60mn + 25n^2 = (6m + 5n)(6m + 5n).$$
$$16x^2 - 24xy + 9y^2 = (4x - 3y)(4x - 3y).$$
$$49t^2 - 4s^2 = (7t + 2s)(7t - 2s).$$

As the figure 1 is the square of 1, we have according to the same rules:

$$x^2 + 2x + 1 = (x + 1)(x + 1).$$
$$x^2 - 2x + 1 = (x - 1)(x - 1).$$
$$x^2 - 1 = (x + 1)(x - 1).$$

4. In order to ascertain whether an expression of three terms can be resolved into factors as indicated in the preceding paragraph, note first whether the coefficients of two of the terms are squares of whole numbers, and if they have like signs. Then see if the letters of these two terms have exponents divisible by 2. If these conditions are complied with, extract the square roots of each of these two terms; then multiply together the square roots thus obtained and double the coefficient of the product. The result should equal the third term of the given expression. If not, the expression cannot be resolved into factors as indicated, or in other words, it is not a perfect square.

Example: Is $4x^2 + 16y^2 - 16xy$ a perfect square, and can it thus be resolved into two like factors?

4 and 16 in the two first terms are squares of whole numbers and have like signs $(+)$; the exponents of the letters in these two terms are divisible by 2. The square root of $4x^2$ is $2x$, and the square root of $16y^2$ *is* $4y$. Multiply these two roots together. Then, $2x \times 4y = 8xy$. Double the coefficient of this product; $2 \times 8xy = 16xy$. This result equals the third term of the given expression, hence it can be resolved into two like factors, which are $(2x - 4y)(2x - 4y)$. The minus sign in the factors is used because the term $16xy$ is preceded by a minus sign.

The difference between two quantities, each of which is a perfect square, can always be resolved into factors, as indicated by the example

$$a^2 - b^2 = (a + b)\ (a - b).$$

It is only necessary to determine if each of the terms is a perfect square. The square roots then take the places of a and b in the sample formula. For example:

$$81x^4y^2 - 64a^2 = (9x^2y + 8a)(9x^2y - 8a).$$

An expression of the form $a^2 + b^2$ cannot be resolved into factors.
A number of examples are given in the following which should be carefully studied. They show the application of the various rules for factoring given in the preceding paragraphs.

Examples:

(1). $6ax^2y^3 - 24ay^7 =$
$6ay^3(x^2 - 4y^4) =$
$6ay^3(x + 2y^2)(x - 2y^2).$

(2). $160\ x^2y^2 - 80xy^2 + 10y^2 =$
$10y^2(16x^2 - 8x + 1) =$
$10y^2(4x - 1)\ (4x - 1).$

(3). $ac + ad - bc - bd =$
$a(c + d) - b(c + d) =$
$(c + d)(a - b).$

(4). $a^2 - ab - bc + ac =$
$a^2 + ac - ab - bc =$
$a(a + c) - b(a + c) =$
$(a - b)(a + c).$

(5). $49m^4n^4 + 14m^2n^2 + 1 =$
$(7m^2n^2 + 1)(7m^2n^2 + 1) = (7m^2n^2 + 1)^2.$

Fractions

A fraction, in algebra, is any expression in which a fraction line is used to indicate a division. In fact, the divison sign is seldom used in algebra, but the dividend is generally written as the numerator and the divisor as the denominator of a fraction. For example, instead of writing $(6a + 3b) \div 5c$, we write:

$$\frac{6a + 3b}{5c}$$

Algebraic fractions can be reduced to their simplest form by cancellation of like factors, the same as in arithmetic. Note that expressions enclosed within parentheses, as $(x + y)$, are to be considered as single letters or symbols in all cases involving cancellation.

Examples:

(1). $\dfrac{6a^2}{3a} = \dfrac{2 \times 3 \times a \times a}{3 \times a} = 2a.$

(2). $\dfrac{6a^2b}{2ac} = \dfrac{6 \times a \times a \times b}{2 \times a \times c} = \dfrac{3ab}{c}.$

(3). $\dfrac{x^2 - y^2}{x - y} = \dfrac{(x + y)(x - y)}{x - y} = x + y.$

(4). $\dfrac{3ax^2 - 3ay^2}{6a(x + y)} = \dfrac{3a(x^2 - y^2)}{6a(x + y)} =$

$\dfrac{3a(x + y)(x - y)}{6a(x + y)} = \dfrac{x - y}{2}.$

Finding the Least Common Denominator

In adding or subtracting fractions, whether in arithmetic or algebra, it is necessary that all the fractions have a common denominator. The least common denominator is found in algebra in a manner similar to that used in arithmetic. Resolve all the denominators into their factors. The least common denominator must contain every type or kind of factor *at least once;* and if any factor occurs in any one denominator more than once, it must be used in the least common denominator *as many times as it occurs in any one of the given denominators.*

Find the least common denominator of the fractions

$$\frac{6a}{2ab^2}, \frac{5b}{3a^2b}, \frac{5ab}{6ac}.$$

Denominators:	$2ab^2$	$3a^2b$	$6ac$.
Factors:	$2 \times a \times b \times b$	$3 \times a \times a \times b$	$2 \times 3 \times a \times c$.

The least common denominator must contain tho factors 2; 3; a (two times); b (two times); and c. Hence, the least common denominator is:

$$2 \times 3 \times a \times a \times b \times b \times c = 6a^2b^2c.$$

Examples:

(1). Denominators: $6a(1-b)$ $(1-b)^2$.
Factors: $6 \times a \times (1-b)$ $(1-b)(1-b)$.
Least common denominator: $6a(1-b)\ (1-b) = 6a(1-b)^2$.

(2). Denominators: $2a^2b^3$ $3ab^3c$
Factors: $2 \times a \times a \times b \times b \times b$ $3 \times a \times b \times b \times b \times c$
Least common denominator: $2 \times 3 \times a \times a \times b \times b \times b \times c = 6a^2b^3c$.

Addition and Subtraction of Fractions

When the method for obtaining the least common denominator is understood, the addition and subtraction of fractions is a simple matter.

After the least common denominator has been found, multiply the numerator of each fraction by the quotient obtained by dividing the least common denominator by the original denominator of each fraction, the same as in arithmetic.

Example:

$$\frac{6a}{2ab^2} + \frac{5b}{3a^2b} + \frac{5ab}{6ac}.$$

The least common denominator is $6a^2b^2c$. The numerator of the first fraction will be multiplied by:

$$\frac{6a^2b^2c}{2ab^2} = 3ac.$$

Hence, $6a \times 3ac = 18a^2c$.

Second numerator is multiplied by:

$$\frac{6a^2b^2c}{3a^2b} = 2bc.$$

Hence, $5b \times 2bc = 10b^2c$.

Third numerator is multiplied by:

$$\frac{6a^2b^2c}{6ac} = ab^2.$$

Hence $5ab \times ab^2 = 5a^2b^3$.

The whole expression reduced to a common denominator is then:

$$\frac{18a^2c + 10b^2c + 5a^2b^3}{6a^2b^2c}$$

In the example given, further simplification is impossible, because the terms in the numerator cannot be added to make the expression more compact. Wherever the terms in the numerator can be added, this is done after the terms are redneed to a common denominator.

Examples:

(1). $\frac{a}{2} + \frac{a}{3} + \frac{1}{a}$.

The least common denominator $= 2 \times 3 \times a = 6a$.

The fractions reduced to the least common denominator are:

$$\frac{3a^2}{6a} + \frac{2a^2}{6a} + \frac{6}{6a} = \frac{5a^2 + 6}{6a}.$$

(2). $\frac{2a - 1}{5} + \frac{3a - 2}{6} + \frac{4a - 3}{3}$.

The least common denominator $= 2 \times 3 \times 5 = 30$. Hence:

$$\frac{12a - 6}{30} + \frac{15a - 10}{30} + \frac{40a - 30}{30} = \frac{67a - 46}{30}.$$

(3). $\frac{3a - 4b}{6} - \frac{3a - 4b}{8} =$

$$\frac{12a - 16b}{24} - \frac{9a - 12b}{24} =$$

$$\frac{12a - 16b - (9a - 12b)}{24} =$$

$$\frac{12a - 16b - 9a + 12b}{24} = \frac{3a - 4b}{24}.$$

Multiplication of Fractions

The rule for multiplication in algebra is identical with the rule used in arithmetic: *The product of the numerators of the factors equals the numerator, and the product of the denominators of the factors equals the denominator of the result.*

Examples:

(1). $\frac{a}{5} \times \frac{b}{6} = \frac{ab}{30}$.

(2). $\frac{a^2b}{c} \times \frac{ab^2}{d} = \frac{a^3b^3}{cd}$.

(3). $\frac{6a}{5} \times \frac{3b}{d} = \frac{18ab}{5d}$.

Cancellation of like factors in numerator and denominator simplifies the calculation, the same as in arithmetic.

(4). $\frac{6a}{b} \times \frac{3b}{c} \times \frac{2c}{9a} = \frac{2 \times 3 \times 6abc}{9abc} = 4$.

(5). $\frac{3xy^2t}{5x^2y} \times \frac{5t}{6xy} = \frac{3 \times 5xy^2t^2}{5 \times 6x^3y^2} = \frac{t^2}{2x^2}$.

(6). $\frac{2(x^2 - 1)}{3} \times \frac{18}{7(x + 1)} = \frac{2 \times 18 \times (x + 1)(x - 1)}{3 \times 7(x + 1)} = \frac{12(x - 1)}{7}$

The last example indicates that quantities consisting of more than one term should be factored, if possible, to permit cancellation.

Division of Fractions

The division of fractions in algebra follows the identical rule used in arithmetic: *Invert the divisor, and proceed as in multiplication.*

Examples:

(1). $\frac{3a}{7} \div \frac{5a}{3} = \frac{3a}{7} \times \frac{3}{5a} = \frac{9a}{35a} = \frac{9}{35}$.

(2). $\frac{6x^3y}{5a^3b} \div \frac{9xy^3}{5a^4b^2} = \frac{6x^3y}{5a^3b} \times \frac{5a^4b^2}{9xy^3} = \frac{5 \times 6x^2ya^4b^2}{5 \times 9xy^3a^3b} = \frac{2xab}{3y^2}$.

(3). $\left(a + \frac{b}{c}\right) \div \frac{a}{b} = \left(\frac{ac}{c} + \frac{b}{c}\right) \div \frac{a}{b} = \frac{ac + b}{c} \times \frac{b}{a} = \frac{(ac + b)b}{ac} = \frac{abc + b^2}{ac}$.

Square Roots

From the rules given for positive and negative quantities we have:

$$(+a) \times (+a) = +a^2, \text{ and } (-a) \times (-a) = +a^2.$$

From this we can draw the conclusion that $\sqrt{a^2}$ may equal either $+a$ or $-a$. In fact, every quantity has algebraically two square roots, numerically equal, but one positive and one negative. The significance of this will be shown in the study of quadratic equations. The fact that a root can be either positive or negative is indicated by using the sign $\pm$. Hence, $\sqrt{4} = \pm 2$.

The square root of an algebraic qnantity is extracted as follows: Extract the square root of the numerical coefficients as in arithmetic, and divide the exponent of each letter under the square root sign by 2. Prefix the sign $\pm$ to the root thus obtained.

Examples:

(1). $\sqrt{16a^2b^2} = \pm 4ab.$

(2). $\sqrt{25a^6b^4c^2} = \pm 5a^3b^2c.$

(3). $\sqrt{\dfrac{36a^4}{b^2}} = \pm \dfrac{6a^2}{b}.$

It is not always possible to extract the square root of the complete quantity beneath the square root sign, and yet retain whole numbers as exponents. In such cases part of the expression may be left under the root sign, while the square root is extracted of the remainder.

Examples:

(1). $\sqrt{16a^2b} = \pm 4a\sqrt{b}.$

(2). $\sqrt{\dfrac{16a^2}{b}} = \sqrt{\dfrac{16a^2 \times b}{b \times b}} = \sqrt{\dfrac{16a^2b}{b^2}} = \pm \dfrac{4a}{b}\sqrt{b}.$

The following methods with regard to the coefficient are also used:

(3). $\sqrt{125a^2b^2} = \sqrt{5 \times 25a^2b^2} = \pm 5ab\sqrt{5}.$

(4). $\sqrt{108a^2b} = \sqrt{3 \times 36a^2b} = \pm 6a\sqrt{3b}.$

CHAPTER IV

SOLVING EQUATIONS OF THE FIRST DEGREE

Examples of simple equations of the first degree and their use have already been given. With the knowledge now acquired of the methods of algebraic operations, we can solve more complicated examples. The general rule for equations of the first degree with one unknown may be comprehensively given as follows:

Clear the equation of all fractions and parentheses. Transpose all the terms containing the unknown x to one side of the equals sign, and all the other terms to the other side. Simplify the expressions as far as possible by addition or subtraction of like terms. Then divide both sides by the coefficient of the unknown x.

The following rules will aid in clearing the fractions and simplifying the equation:

1. When all the terms in an equation have been reduced to the same denominator, this denominator can be canceled in all the terms.

Example:

$$\frac{5}{x-1}+\frac{2x}{x-1}=\frac{3x}{x-1}.$$

Cancel the denominator $x-1$, because it appears in *all* the terms. Then:

$$5+2x=3x;\ 5=3x-2x;\ 5=x.$$

2. A term which divides *all* the terms on one side of the equals sign may be transposed to the other side, if it is made to multiply *all* the terms on that side.

Examples:

(1). $\frac{5}{2x}+\frac{3}{2x}=9.$

$$5+3=9\times 2x;\ 8=18x,\ x=\frac{8}{18}=\frac{4}{9}.$$

(2). $\frac{3x}{5}+\frac{9x+1}{5}=9+x.$

$$3x+9x+1=5\times 9+5x;\ 7x=44;\ x=6\frac{2}{7}.$$

3. A term which multiplies *all* the terms on one side of the equals sign may be transposed to the other side, if it is made to divide *all* the terms on that side.

Examples:

(1). $5(x+1)+5(x+2)=15+25x.$

$$(x+1)+(x+2)=\frac{15}{5}+\frac{25x}{5}.$$

$$x+1+x+2=3+5x;\ 0=3x;\ x=0.$$

(2). $2x(x-3)=4x-2x^2.$

$$x-3=\frac{4x}{2x}-\frac{2x^2}{2x}.$$

$$x-3=2-x;\ 2x=5;\ x=2\tfrac{1}{2}.$$

The rule that any term preceded by a + sign on one side of the equals sign may be transposed to the other side if the sign is changed to —, and that any term preceded by a — sign may be transposed to the other side if the sign is changed to +, will be remembered from the paragraphs on "Transposition," in the first part of this book. By means of the rules given, the following examples can be solved. Always remember that the object of all operations with an equation is to *obtain the unknown x on one side of the equals sign, with a coefficient equal to 1 (that is, apparently without a coefficient), and all the known quantities on the other side.*

Examples:

(1). $3x+2x-4=21.$
$3x+2x=21+4.$
$5x=25.$
$x=5.$

(2). $15x=57-9x+5x.$
$15x+9x-5x=57.$
$19x=57.$
$x=3.$

(3). $19x - 17 - 3x + 72 = 12 + 9x - 25 + 4x + 83.$
$19x - 3x - 9x - 4x = 12 - 25 + 83 + 17 - 72.$
$3x = 15.$
$x = 5.$

(4). $36 = -9x.$
$\frac{36}{-9} = x.$
$-4 = x.$

(5). $6x = -4.$
$x = \frac{-4}{6}.$
$x = -\frac{2}{3}.$

(6). $\frac{x}{5} = 8.$
$\frac{5x}{5} = 5 \times 8.$
$x = 40$

(7). $5 \div x = 8.$
$\frac{5}{x} = 8.$
$\frac{5x}{x} = 8x.$
$5 = 8x.$
$5/8 = x.$

(8). $7(x - 2) = 35.$
$7x - 14 = 35.$
$7x = 35 + 14.$
$7x = 49.$
$x = 7.$

(9). $b(x - a) = c.$
$bx - ab = c.$
$bx = c + ab.$
$x = \frac{c + ab}{b}.$

Note: When letters are used to represent *known* quantities, they are usually selected at the beginning of the alphabet, as a, b, c, etc.

(10). $9(x + 2) - 3(x - 2) = 12(x - 4).$
$9x + 18 - 3x + 6 = 12x - 48.$
$18 + 6 + 48 = 12x - 9x + 3x.$
$72 = 6x;\ 13 = x.$

(11). $\frac{51}{5 - x} - 3 = 14.$
$\frac{51}{5 - x} - \frac{3(5 - x)}{5 - x} = \frac{14(5 - x)}{5 - x}.$
$51 - 3(5 - x) = 14(5 - x).$
$51 - 15 + 3x = 70 - 14x.$
$3x + 14x = 70 - 51 + 15.$
$17x = 34;\ x = 2.$

(12). $\frac{5}{x - 3} = \frac{3}{5 - x}.$
$\frac{5(5 - x)}{(x - 3)(5 - x)} = \frac{3(x - 3)}{(5 - x)(x - 3)}.$
$5(5 - x) = 3(x - 3).$
$25 - 5x = 3x - 9.$
$25 + 9 = 3x + 5x.$
$34 = 8x;\ 4\frac{1}{4} = x.$

Examples for Practice

(1). $3 = x - 7$. Answer: $x = 10$.

(2). $7x = -4.2$. Answer: $x = -0.6$.

(3). $9x + 4 = 31$. Answer: $x = 3$.

(4). $8x = 11 - 5x + 2x$. Answer: $x = 1$.

(5). $10x + 29 = 13 + 17x + 18 - 2x - 97$. Answer: $x = 19$.

(6). $7 = \frac{x}{3}$. Answer: $x = 21$.

(7). $16 = x \div \frac{5}{8}$. Answer: $x = 10$.

In the example above, remember that $x \div \frac{5}{8} = x \times \frac{8}{5} = \frac{8}{5}x$. Hence $\frac{8}{5}$ is the coefficient of x.

(8). $5 - \frac{x}{7} = 4$. Answer: $x = 7$.

(9). $6(x - 9) = 27$. Answer: $x = 13.5$.

(10). $a(a + x) = b$. Answer: $x = \frac{b - a^2}{a}$.

(11). $5(x + 1) - 2(3x - 1) = x$. Answer: $x = 3\frac{1}{2}$.

(12). $\frac{x}{3} + \frac{x}{4} = 14$. Answer: $x = 24$.

(13). $\frac{5x - 4}{4x - 5} = \frac{4}{5}$. Answer: $x = 0$.

(14). $\frac{87}{3x - 1} + 17 = 20$. Answer: $x = 10$.

Problems leading to Equations with One Unknown Quantity

1. A certain work is to be done by three men, A, B and C. Alone, A could do the work in 10 hours, B in 12 hours, and C in 15 hours. How long a time will be required when they work together?

Total number of hours required $= x$.

A completes in one hour 1/10 of the work; thus, in x hours, $1/10x$.

Similarly, B completes in x hours $1/12x$, and C, $1/15x$ of the work; but their combined work completes the job. Hence,

$$\frac{1}{10}x + \frac{1}{12}x + \frac{1}{15}x = 1.$$

$$\frac{6}{60}x + \frac{5}{60}x + \frac{4}{60}x = 1.$$

$$6x + 5x + 4x = 60.$$

$$15x = 60.$$

$$x = 4 \text{ hours.}$$

2. In a triangle, the sum of the lengths of the three sides is 17 feet. The first side is 2 feet shorter than the second and ¾ of the length of the third. Find the length of the sides.

Length of third side $= x$.

Length of first side $= \frac{3}{4}x$.

Length of second side $= \frac{3}{4}x + 2$.

Then,

$$x + \frac{3}{4}x + \frac{3}{4}x + 2 = 17.$$

$$2\tfrac{1}{2}x = 17 - 2 = 15.$$

$$x = 6.$$

The third side thus equals 6 feet.

The first side equals $\frac{3}{4}x = \frac{3}{4} \times 6 = 4\tfrac{1}{2}$ feet.

The second side equals $\frac{3}{4}x + 2 = 4\tfrac{1}{2} + 2 = 6\tfrac{1}{2}$ feet.

Equations with Two Unknown Quantities

When an equation contains two unknown quantities, the values of these quantities can be determined only if *two* independent equations are given, each containing the unknown quantities. It is common practice to call the unknown quantities x and y.

Equations with two unknown quantities are solved by so combining them that an equation with one unknown is found, which can be solved by the methods already explained.

Solve x and y in the equations:

$$5x + 3y = 22.$$

$$3x - y = 2.$$

In one of the equations, solve the value of one of the unknown, in terms of the other. Insert this value in the other equation, thus obtaining an equation with one unknown.

Following this rule, solve for y in the second equation, which is evidently the simplest method, as y in this equation has no coefficient (or, rather, a coefficient equal to 1).

$$3x - y = 2.$$

$$3x - 2 = y.$$

Insert this value of y in place of y in the first equation:

$$5x + 3(3x - 2) = 22.$$

$$5x + 9x - 6 = 22.$$

$$14x = 28.$$

$$x = 2.$$

Now insert this value of x in the expression for y found above:

$$3x - 2 = y.$$

$$3 \times 2 - 2 = y.$$

$$4 = y.$$

Hence, $x = 2$, and $y = 4$.

To prove the calculation, insert the values found in one of the original equations:

$$5x + 3y = 22.$$
$$5 \times 2 + 3 \times 4 = 22.$$
$$10 + 12 = 22.$$

This proves that the values found are correct, and that no error has been made in the calculations.

Another method is as follows: *Solve for one of the unknown in both equations, and place the quantities thus found in a new equation, from which the other unknown can now be solved.*

Apply this rule to the same equations as before.

$$5x + 3y = 22.$$
$$3x - y = 2.$$

Solve for y in the first equation:

$$5x + 3y = 22.$$
$$3y = 22 - 5x.$$
$$y = \frac{22 - 5x}{3}.$$

Now solve for y in the second equation:

$$3x - y = 2.$$
$$3x - 2 = y.$$

Here we have now two expressions both of which are equal to y. Hence, these expressions must themselves be equal:

$$3x - 2 = \frac{22 - 5x}{3}.$$

In this equation solve for x.

$$3(3x - 2) = 22 - 5x.$$
$$9x - 6 = 22 - 5x.$$
$$9x + 5x = 22 + 6.$$
$$14x = 28.$$
$$x = 2.$$

Then, $y = 4$, is found as in the preceding case.

There is still a third method that may often be used to advantage. *Select the unknown quantity to be eliminated from both equations. Multiply one equation by such a factor that the coefficients of the unknown to be eliminated will be numerically equal. If the unknown terms to be eliminated have like signs in the two equations, subtract one equation from the other; if they have unlike signs, add one to the other.*

$$5x + 3y = 22.$$
$$3x - y = 2.$$

Select y as the unknown to be eliminated, as it will be seen that by multiplying the second equation by 3, the y-terms will then have numerically equal coefficients. Multiply the second equation by 3:

$$3 \times 3x - 3 \times y = 2 \times 3.$$
$$9x - 3y = 6.$$

Now place this equation beneath the first, and add the two together.

$$\begin{array}{l} 5x + 3y = 22. \\ 9x - 3y = 6. \\ \hline 14x \qquad = 28. \\ \qquad x = 2. \end{array}$$

The unknown y is now found in the same manner as before, by inserting the value of x just found into any of the given equations and solving for y.

$$5 \times 2 + 3y = 22.$$
$$3y = 22 - 10.$$
$$3y = 12.$$
$$y = 4.$$

Examples for Practice

Some of the examples in the following are partly worked out to indicate the method used.

(1). $x + y = 15.$
$x - y = 6.$

$2x \qquad = 21.$
$x = 10.5.$

$10.5 + y = 15.$
$y = 15 - 10.5.$
$y = 4.5.$

(2). $x + 3y = 20.$
$2x - 5y = 29.$
Then:
$2x + 6y = 40.$
$2x - 5y = 29.$

$11y = 11.$
$y = 1;\ x = 17.$

(3). $5(x - 2) - 3(y - 1) = 0.$
$2(x - 2) + 7(y - 1) = 0.$

Mnltiply, to eliminate parentheses:

$5x - 10 - 3y + 3 = 0.$
$2x - 4 + 7y - 7 = 0.$

Transpose unknown quantities to one side:

$5x - 3y = 10 - 3 = 7.$
$2x + 7y = 4 + 7 = 11.$

Answer: $x = 2;\ y = 1.$

(4). $x + 36y = 450.$
$36x + y = 660.$

Answer: $x = 18;\ y = 12$

(5). $1.5x + 3.5y = 33.$
$x + 2y = 20.$

Answer: $x = 8; y = 6$

(6). $\frac{6}{x} - \frac{3}{y} = 4.$

$\frac{8}{x} + \frac{15}{y} = -1.$

Answer: $x = 2;\ y = -3.$

(7). $\frac{3}{x}+\frac{5}{y}=2.$

$\frac{9}{x}-\frac{10}{y}=1.$

In this case apply the method of adding the two equations, it being easily seen by inspection that the y-terms can be made alike by multiplying the first eqnation by 2. Hence, multiply tbe first equation by 2, and add the equations:

$$\frac{6}{x}+\frac{10}{y}=4.$$

$$\frac{9}{x}-\frac{10}{y}=1.$$

$$\frac{6}{x}+\frac{9}{x}=5.$$

Clearing the fractions:

$$6+9=5x;\ 15=5x;\ x=3, \text{ and } y=5.$$

Problems Leading to Equations with Two Unknown Quantities

1. The sum of two numbers is 1000. The difference between the numbers is 222. Find the numbers.

The numbers to be found are x and y.

$$x+y=1000.$$
$$x-y=\ \ 222.$$

Answer: $x=611$; $y=389$.

2. A is 27 years older tban B. Ten years ago he was 10 times as old as B. How old are A and B?

A is x years old; B, y years. Hence:

$$x-27=y.$$

But 10 years ago, A's age was $(x-10)$, and B's, $(y-10)$. Hence:

$$x-10=10\times(y-10).$$

From these two equations we find $x=40$; $y=13$.

3. A tank of 120 gallons capacity can be filled from two pipes; if one pipe is open for 6 minutes, and the other for 3 minutes, 100 gallons will enter the tank. If the first pipe is open 3 minutes and the second 6 minutes, 110 gallons will enter the tank. How long a timo would be required for each of the pipes to fill the tank?

One pipe requires x minutes, the other y minutes.

In one minute the first pipe delivers $\frac{120}{x}$ gallons.

In one minute the second pipe delivers $\frac{120}{y}$ gallons.

Hence:

$$\frac{120}{x} \times 6 + \frac{120}{y} \times 3 = 100.$$

$$\frac{120}{x} \times 3 + \frac{120}{y} \times 6 = 110.$$

Multiply the first equation by 2, and subtract the second equation from the first:

$$\frac{120 \times 12}{x} + \frac{120 \times 6}{y} = 200.$$

$$\frac{120 \times 3}{x} + \frac{120 \times 6}{y} = 110.$$

$$\frac{120 \times 12}{x} - \frac{120 \times 3}{x} = 90.$$

$$1440 - 360 = 90x; \quad x = 12; \quad y = 9.$$

CHAPTER V

SOLVING QUADRATIC EQUATIONS

A *quadratic* equation is one in which the unknown quantity is contained in the second, or first and second, power. The following equations are, therefore, examples of quadratic equations:

$$x^2 - 27 = 39.$$

$$3x^2 - 5x = 12.$$

A quadratic equation is frequently called an equation of the *second degree.*

A *pure quadratic* equation is one which contains the unknown in the second power only, as $x^2 - 12 = 4$.

An *affected quadratic* equation is one which contains the unknown in both the first and second power, as $x^2 - 3x = 4$.

Solving Pure Quadratic Equations

A pure quadratic equation can always be simplified so that it takes the form $x^2 = a$; that is, the unknown quantity in the second power, without a coefficient, will be on one side of the equals sign, and the known quantities, reduced to their simplest form, will be on the other side.

To solve the equation when in this form, it is only necessary to extract the square root on each side of the equals sign. Remember that roots have both positive and negative signs, or that the square root of $a^2 = \pm a$.

Examples:

(1). $x^2 = 16.$

$\sqrt{x^2} = \sqrt{16}.$

$x = \pm 4.$

(2). $3x^2 = 75.$

$x^2 = 25.$

$x = \pm 5.$

(3). $7x^2 - 8 = 9x^2 - 10.$

$10 - 8 = 9x^2 - 7x^2.$

$2 = 2x^2.$

$1 = x^2.$

$\pm 1 = x.$

(4). $(3x - 4)(3x + 4) = 65.$

Multiply and reduce:

$9x^2 - 16 = 65.$

$9x^2 = 81.$

$x^2 = 9.$

$x = \pm 3.$

(5). $(7 + x)^2 + (7 - x)^2 = 130.$

$(7 + x)(7 + x) + (7 - x)(7 - x) = 130.$

Multiplying, we have:

$49 + 14x + x^2 + 49 - 14x + x^2 = 130.$

$2x^2 = 130 - 49 - 49 = 32.$

$x^2 = 16.$

$x = \pm 4.$

Solving Affected Quadratic Equations

An affected quadratic equation can be reduced to the form $x^2 + px = q$; that is, the unknown qnantity in the second power, has a coefficient equal to 1; the unknown in the first power has a known coefficient p; and q represents the known quantities reduced to their simplest form. The equation $x^2 - 3x = 4$ is an example of an equation reduced to the form $x^2 + px = q$. The coefficient of x in this case is (-3); hence the minus sign.

To reduce an equation to its simplest form, add all the like terms together; arrange them in the form given above, and then divide by the coefficient of x^2.

Examples:

(1). Reduce $6x^2 + 7 = 3x - x^2 + 12$ to its simplest form.

$$7x^2 - 3x = 5.$$

$$x^2 - \tfrac{3}{7}x = \tfrac{5}{7}$$

(2). Reduce $\frac{1}{2}x^2 - 2x + 13 = 0$ to its simplest form.

$$\tfrac{1}{2}x^2 - 2x = -13.$$

$$x^2 - \frac{2}{\frac{1}{2}}x = -\frac{13}{\frac{1}{2}}.$$

$$x^2 - 4x = -26.$$

When the equation has been reduced to the form $x^2 + px = q$, the value of x may be found by the formula:

$$x = -\frac{p}{2} \pm \sqrt{\frac{p^2}{4} + q}.$$

It is not necessary, in order to make use of this formula, to know how it is obtained. Equations can be solved by its use quickly and with the minimum risk of error. For the benefit of those students who wish to know how the formula is obtained, this will be explained later. This formula should be committed to memory, as it is applied to all quadratic equations. Expressed as a rule the formula says: *x equals half the coefficient for the second term (p) with changed sign, plus or minus the square root of the square of half the coefficient of the second term, plus the known quantity (q).*

Remember that if the known quantity is negative, $+(-q) = -q$.

Examples:

(1). $x^2 + 3x = 28.$

$$x = -\frac{3}{2} \pm \sqrt{\left(\frac{3}{2}\right)^2 + 28} = -\frac{3}{2} \pm \sqrt{\frac{9}{4} + 28.}$$

$$x = -\frac{3}{2} \pm \sqrt{\frac{9+112}{4}} = -\frac{3}{2} \pm \frac{11}{2}.$$

$$x = -\frac{3}{2} + \frac{11}{2} = \frac{8}{2} = 4$$

$$\text{or } x = -\frac{3}{2} - \frac{11}{2} = -\frac{14}{2} = -7.$$

The two values of x found are obtained by using either the plus or minus sign of the root. Either value, if inserted in the original equation, will *satisfy* it.

$$x^2 + 3x = 28.$$

Insert $x = 4.$

$16 + 12 = 28.$

Insert $x = -7.$

$49 + 3 \times (-7) = 28.$

$49 - 21 = 28.$

(2). $x^2 - 4x = 21.$

$x = +2 \pm \sqrt{(2)^2 + 21} = 2 \pm \sqrt{25}.$

$x = 2 \pm 5.$

$x = 7$, or $x = -3.$

It is common practice to indicate the two roots of the equation x_1 and x_2, respectively. The small numbers ($_1$) and ($_2$) are not exponents, when placed at the lower corner of the letter. Hence, we would write,

$$x_1 = 7, \text{ and } x_2 = -3.$$

(3). $3x^2 - 24x + 45 = 0.$

$3x^2 - 24x = -45.$

Divide the terms by 3, the coefficient of x^2:

$x^2 - 8x = -15.$

$x = +4 \pm \sqrt{(4)^2 - 15} = +4 \pm \sqrt{16 - 15}.$

$x = +4 \pm 1.$

$x_1 = 5;\ x_2 = 3.$

(4). $(x-3)(x-5)=0.$
Multiply:
$x^2-3x-5x+15=0.$
Reduce and arrange for solving:
$x^2-8x=-15.$
$x=+4\pm\sqrt{16-15}.$
$x=4\pm1.$
$x_1=5;\ x_2=3.$

Examples for Practice

(1). $x^2=81.$	Answer: $x=\pm9.$
(2). $5x^2=125.$	Answer: $x=\pm5.$
(3). $9x^2-36=5x^2.$	Answer: $x=\pm3.$
(4). $7x^2-8=24-x^2.$	Answer: $x=\pm2.$
(5). $x^2-8x=20.$	Answer: $x_1=10;\ x_2=-2.$
(6). $x^2-8x=-12.$	Answer: $x_1=6;\ x_2=2.$
(7). $x^2-x=12.$	Answer: $x_1=4;\ x_2=-3.$
(8). $6x^2+48x=54.$	Answer: $x_1=1;\ x_2=-9.$

Problems Leading to Quadratic Equations

1. The sides enclosing the right angle in a right-angled triangle are in the proportion 12:5. The side opposite the right angle is 299 feet long. Find the length of the other sides.

In a right-angled triangle, the square of the side opposite the right angle equals the sum of the squares of the sides enclosing this angle.

One of the sides to be found $=x$; the other side to be found $=\frac{12}{5}x$.

Then:

$$x^2+\left(\frac{12}{5}x\right)^2=299^2.$$

Use tables in an engineering handbook to obtain squares and square roots of large numbers.

$$x^2+\frac{144}{25}x^2=89{,}401.$$
$$25x^2+144x^2=89{,}401\times25.$$
$$169x^2=2{,}235{,}025.$$
$$x^2=\frac{2{,}235{,}025}{169}=13{,}225.$$
$$x=115\text{ feet.}$$

In this case we do not use the negative value of the root, because, while it satisfies the *equation*, it does not apply to the *problem* here presented. We could not conceive of a side -115 feet long.

The other side then equals $\frac{12}{5}\times115=276$ feet.

2. Find two factors of 96 the sum of the squares of which is 208.

The factors are x and $\frac{96}{x}$. Note that $x\times\frac{96}{x}=96.$

$$x^2 + \left(\frac{96}{x}\right)^2 = 208$$

Note that $\left(\frac{96}{x}\right)^2 = \frac{96^2}{x^2}$.

$$\frac{x^4}{x^2} + \frac{9216}{x^2} = \frac{208x^2}{x^2}.$$

$$x^4 - 208x^2 = -9216.$$

An eqnation having the unknown in the fourth and second power only can be solved just as one having the unknown in the second and first powers; only, in this case, x^2 takes the place of x in the formula for solving the unknown.

$$x^2 = +104 \pm \sqrt{104^2 - 9216}.$$

$$x^2 = +104 \pm \sqrt{10{,}816 - 9216} = 104 \pm 40.$$

$$x_1^2 = 144, \text{ and } x_2^2 = 64.$$

Now extract the square roots of $x^2 = 144$ and $x^2 = 64$.

Then, $x_1 = \pm 12$, and $x_2 = \pm 8$.

Here both roots can be used because $12 \times 8 = 96$, and $-12 \times -8 = 96$.

3. A certain number of men agreed to pay an equal share for the purpose of huyiug a machine that was to cost $6300. Two of the men later changed their mind, and the remainder each put up $200 more than originally agreed in order to make up for the shares of these two. How many men made the first agreement?

The original number of men $= x$.

The share of each, originally $= \frac{6300}{x}$.

The number of meu actually buying the machine $= x - 2$.

The share of each of these $= \frac{6300}{x-2}$.

This last share was $200 greater than the original amounts assessed; hence:

$$\frac{6300}{x-2} - \frac{6300}{x} = 200.$$

$$6300x - 6300\,(x-2) = 200x(x-2).$$

$$6300x - 6300x + 12{,}600 = 200x^2 - 400x.$$

$$12{,}600 = 200x^2 - 400x.$$

Divide by 200, and arrange equation like sample form:

$$x^2 - 2x = 63.$$

$$x = +1 \pm \sqrt{1 + 63} = +1 \pm 8.$$

$$x_1 = 9;\ x_2 = -7.$$

The number of men was 9. The root — 7 is impossible as a solution of the problem, although it mathematically satisfies the equation.

Deducing the Formula for x in a Quadratic Equation

The method by means of which the formula

$$x = -\frac{p}{2} \pm \sqrt{\frac{p^2}{4} + q}$$

is obtained from the equation $x^2 + px = q$ is as follows:

Add to the left-hand side of the equation a quantity such that this side will be a perfect square of the form $a^2 + 2ab + b^2$. The added quantity must, of course, also be added to the right side of the equation in order to retain the relation of equality.

The quantity to be added is $\frac{p^2}{4}$.

$$x^2 + px + \frac{p^2}{4} = \frac{p^2}{4} + q.$$

Now, as $\left(x + \frac{p}{2}\right)\left(x + \frac{p}{2}\right) = x^2 + px + \frac{p^2}{4}$, we can extract the sqnare root on both sides of the equals sign:

$$\sqrt{x^2 + px + \frac{p^2}{4}} = \sqrt{\frac{p^2}{4} + q}.$$

$$x + \frac{p}{2} = \pm\sqrt{\frac{p^2}{4} + q}.$$

$$x = -\frac{p}{2} \pm \sqrt{\frac{p^2}{4} + q}.$$

If the given equation is of the form

$$ax^2 + bx = c,$$

then the unknown qnantity x may be found directly by the formula:

$$x = \frac{-b \pm \sqrt{b^2 + 4ac}}{2a}.$$

Note that the sign of the coefficient for x in the first power (b) changes its sign, the same as in the formula already given.

Example:

$$3x^2 - 15x = 0.$$

$$x = \frac{+15 \pm \sqrt{225 + 0}}{2 \times 3} = \frac{15 \pm 15}{6}$$

$$x_1 = \frac{30}{6} = 5; \quad x_2 = \frac{0}{6} = 0.$$